FLORA OF TROPICAL EAST AFRICA

CANNABACEAE*

B. VERDCOURT

Annual or perennial erect or climbing herbs without latex. Leaves alternate or opposite, simple, undivided or palmately lobed or divided into separate leaflets; petioles well developed; stipules present, free or fused. Flowers mostly dioecious, axillary, wind-pollinated. Male paniculate; perianth 5-partite with imbricate segments; stamens 5, opposite the tepals, the anthers straight, erect in bud, 2-thecous, at first dehiscing by apical oval pores but soon dehiscing lengthwise; vestigial ovary absent. Female ± sessile, crowded or strobilate, tightly covered or loosely subtended by small or large conspicuous persistent bracteoles; bracts also present; perianth membranous, entire, investing the ovary; ovary superior, sessile, 1-locular with 1 pendulous anatropous ovule; style terminal, short, with 2 long filiform stigmas. Fruit an achene covered by the persistent perianth; endosperm sparse, fleshy and oily; embryo curved or spirally coiled.

A family of only 2 genera occurring naturally in the temperate parts of the northern hemisphere, formerly included in either *Urticaceae* or *Moraceae*. The number of species is in dispute; *Cannabis* is usually considered to be monotypic but a second species *C. ruderalis* Janisch has been described; *Humulus* is usually considered to contain 2 species although 3–4 have been claimed. Although small the family is of considerable economic importance, *Humulus* being used to flavour beer and *Cannabis* to produce the fibre hemp and the drug variously known as bhang, dagga, hashish, pot, marijuana and Indian hemp. A very full account of the family and generic characteristics, together with an extensive bibliography, is given by N. G. Miller (Journ. Arn. Arb. 51: 185–203 (1970)).

CANNABIS

L., Sp. Pl.: 1027 (1753) & Gen. Pl., ed. 5: 453 (1754)

Erect annual aromatic herb. Leaves alternate, or opposite at the base of the stems, palmately divided into narrow leaflets; stipules linear, acute. Male flowers numerous, shortly pedicellate, in lax drooping terminal and axillary panicles; tepals free, oblong, spreading or reflexed; anthers at length pendulous. Female flowers fewer, in short erect spikes, leafy below, each flower sessile in the axil of an enveloping bracteole and also with a stipule-like bract; perianth inconspicuous, very thin, closely enveloping the ovoid ovary. Fruit ovoid, compressed with seed closely conforming to the thin crustaceous pericarp. Embryo strongly curved; cotyledons fleshy.

C. sativa *L.*, Sp. Pl.: 1027 (1753); P.O.A. C: 162 (1895); E.M. 1 (Moraceae): 44 (1898); Hiern, Cat. Afr. Pl. Welw. 1: 994 (1900); Rendle in F.T.A. 6(2): 16 (1916); Hauman in F.C.B. 1: 176 (1948); U.O.P.Z.: 169 (1949); F.P.S. 2: 280 (1952); Trease, Textbook of Pharmacognosy: 216,

---

* Almost invariably spelt as *Cannabinaceae* in older works, also as *Cannabiaceae* and *Cannabidaceae* (used in Willis, Dict. Fl. Pl., ed. 8: 197 (1973)).

FIG. 1. *CANNABIS SATIVA*—**1,** male flowering shoot, × ⅔; **2,** male inflorescence, × 3; **3,** male flower, × 6; **4,** stamen, × 6; **5,** female inflorescence, × 4; **6,** female flower, with bracteole, × 6; **7,** female flower, × 6; **8,** fruit, enveloped by bracteole, × 4; **9,** achene, × 4. 1, from *Ward* 6086; 2–4, from *Semsei* 1667; 5–7, from *Holst* 2685; 8, 9, from *Kennedy* in *F.D.* 1266. Drawn by Miss V. Goaman.

fig. 75 (1952) [floral structure]; E.P.A.: 17 (1953); F.W.T.A., ed. 2, 1: 623 (1958); Schreiber in Hegi, Illustr. Fl. Mitt.-Eur., ed. 2, 3(1): 290–295, fig. 136–7, t. 88/1 (1958) [very full account]; Purseglove, Tropical Crops, Dicotyledons 1: 40, fig. 4 (1968); Verdc. & Trump, Common Poisonous Pl. E. Afr.: 96 (1969); Miller in Journ. Arn. Arb. 51: 188, etc. (1970); Stearn in Joyce & Curry (eds.), The Botany and Chemistry of Cannabis: 1–10, fig. 1–7 (1970). Type: ♀ specimen in *Hort. Cliff.* (BM, lecto.!)

A rough, rather rank-smelling, leafy, simple or branched herb 0·9–4·5 m. tall, the male plants taller and more slender, dying soon after flowering, the females stockier and living several months after pollination; stems angular, covered with rather short stiffish hairs. Leaves 3–7(–11)-foliolate; petioles 3–7·5 cm. long, pubescent; leaflets sessile, narrowly lanceolate, 2·5–15 cm. long, 0·35–2 cm. wide, tapering-acuminate at the apex, narrowly cuneate at the base, the margins coarsely toothed, covered on both surfaces with very short bristly hairs and small yellow glands. Male and female flowers on different plants or rarely on one plant but then one sex predominating. Male inflorescence loosely paniculate, up to 18 cm. long, covered with minute bristly hairs; flowers whitish to yellowish-green; sepals oblong-elliptic, 2·8–4 mm. long, 1–1·6 mm. wide, minutely hairy; stamens at length pendulous, the filaments about 0·3–1 mm. long and the anthers 3–3·7 mm. long. Female inflorescences not projecting from the leaves, more compact, short and few-flowered. Bracteole green, acuminate, enwrapping the ovary and forming a basally swollen tubular sheath 1·8–8·5 mm. long, up to 6 mm. wide opened out, covered with slender hairs and short-stalked or stalkless circular resin-secreting glands. Ovary ± globose, ± 1–1·2 mm. in diameter; stigmas slender, 1·2–7 mm. long, deciduous. Achene ("seed", actually consisting of the seed with a hard shell tightly covered by the thin ovary wall) ellipsoid or subglobose, slightly compressed, keeled, shiny, 2·5–5 mm. long, 2–3·5 mm. wide, greyish to brownish and usually covered with a pale "map-like" network. Fig. 1.

UGANDA. Ankole District: Mbarara, July 1957, *Bushara* in *Lind* 2308; Busoga District: Lolui I., 17 Nov. 1964, *G. Jackson* 12643; Masaka District: 2·5 km. W. of Bunado, near Kisasa, 22 May 1972, *Lye* 6945!
KENYA. Nandi Country, Sibu, *Evan James*!; Kavirondo, *Scott Elliott* 7051!
TANZANIA. Tanga District: Amboni, June 1893, *Holst* 2685!; Kahama/Tabora Districts: about 90 km. NNW. of Tabora, Mininga, Apr. 1867, *Grant* 75!; Njombe District: Ndapo (Mdapo), Mar. 1954, *Semsei* 1667!
DISTR. U2–4; K3–5; T1, 3, 4, 6, 7; Z; P; a central Asian plant now widely cultivated and naturalized throughout the world
HAB. Widely and illegally cultivated as a drug plant, also frequent as a weed of cultivation; 0–2100 m.

SYN. *C. indica* Lam., Encycl. 1: 695 (1785). Type: India, *Sonnerat* (P, syn.)

INDEX TO CANNABACEAE